YOUR KNOWLEDGE HAS VALUE

Bibliographic information published by the German National Library:

The German National Library lists this publication in the National Bibliography; detailed bibliographic data are available on the Internet at http://dnb.dnb.de .

Imprint:

Copyright © 2016 GRIN Verlag, Open Publishing GmbH
Print and binding: Books on Demand GmbH, Norderstedt Germany
ISBN: 9783668305632

This book at GRIN:

http://www.grin.com/en/e-book/340686/bwondha-hailstorm-disaster-assessment-report

Paul Waluube

Bwondha hailstorm disaster assessment report

GRIN Publishing

GRIN - Your knowledge has value

Since its foundation in 1998, GRIN has specialized in publishing academic texts by students, college teachers and other academics as e-book and printed book. The website www.grin.com is an ideal platform for presenting term papers, final papers, scientific essays, dissertations and specialist books.

Visit us on the internet:

http://www.grin.com/

http://www.facebook.com/grincom

http://www.twitter.com/grin_com

Bwondha hailstorm disaster assessment report

July 30, 2016

Introduction

Bwondha by and large is a gazetted landing site along the shores of Lake Victoria in South Busoga region within Malongo Sub County nearby the forest reserve. The community is highly populated because of the economic activities that are going on at the site relating to the water body and historical south Busoga forest reserve.

According to the 2014 Uganda population and Housing census, Malongo Sub County emerged as the most populated sub county in the district with a total population of 102,524 persons. Of this population, approximately 45% are residing in Bwondha parish and most of whom are staying on Bwondha landing site ($\approx$ 35%).

Assessment methodology

1. Discussions

The method of assessment was that upon reaching Bwondha landing site, I commenced with a rapid assessment involving discussion with the sub county chief about the climate change effect. After the discussion with the sub county chief, a focus group discussion was held with the local leaders from the 8 LC I zones and during this discussion, the local leaders revealed that they are aware of the causes of the disaster at the landing site.

The local leaders with whom the discussion was conducted are: Mr. Muhamad Okello, the chair of Mirembe zone, Mr. Muhwana Badru, the Chair of Nalubabwe zone, Mr. Tizoomu Fazir, the chair of Musoma zone, Mr. Hiisa Barrack, the chair of Zone C, Mr. Batambuze Charles, the chair of Makonko zone, Mr. Kunya David, the parish chair for Bwondha, Mr. Irumba Hakim, the Zone B chair, Mr. Kalulu Twaha, the V/chair of zone A and Mr. Ndyeku Rashid, the chair of Zone D.

Results

From the discussions held, the leaders informed the technical officer that the causes of this disaster was broadly categorized as a climate change effect characterized by;

- Too much dry spell whenever it goes beyond March 15 & September 15 of every year.
- Deforestation along the shores of Lake Victoria for instance Lubango, Walujjo, Nairobi, Bwagu mainland. The majority of the leaders attested that there has been

significant destruction of the forest cover over the last 30 years that used to surround the lake along its vicinity. This led to direct exposure of the Bwondha beach community which is just on the shore.

On a micro scale, the cause of the disaster was a heavy rainstorm that occurred at around 11:00 pm leading to breaking of most temporary and semi-permanent buildings most which were residential as well crop stands (bananas and cassava inclusive).

Trends of hazards/disaster occurrence at Bwondha

According to the local inhabitants at the landing site, Bwondha similar hazards occurred in 1994, 2001, 2007, and 2013. Therefore, this was not the first time such an incident occurred in the area.

Trends of settlement of affected communities

The field based interviews conducted during the survey showed that most of the residents on the landing site most of whom are mid age and elder residents (table 1) were migrants from other areas mainly the current Iganga, Namutumba, and Bugwere sub region districts. The majority of the affected households were established by people who migrated to this area at varying periods as showed in table 2.

Table 1: Time of settlement at Bwondha landing site

Period of migration	% age	Age of respondents	%age
Within 5 years	20	Below 24 years	8
5 to 10 years	40	25 to 34 years	24
11 to 20 years ago	32	Above 34 years	68
Born in the sub county	8		

Findings on the nature of effect

There are many effects that have been suffered by Bwondha community following the occurrence of this disaster but some of those that can be reported easily are;

1. Houses were destroyed (approximately 75% of the reported cases) completely destroyed to ground level. My assessment of most of the affected housing structures were built in a semi-permanent state characterized by burnt bricks joined using mud as depicted in figure 1.

Fig. 1a: destroyed residential building

Fig. 1b: devastated residence

2. Some of the dilapidated buildings were built using burnt & un-burnt bricks joined by mud (fig.2). This prompted me to ask the respondents whether they temporary residents since most of them had migrated to this site over the last 20+ years. To my surprise the respondents said these were permanent residences, but there was no correlation of this with the type of buildings built by these residents.

Fig. 1c: fallen residential building

Fig.2a: Completely ravaged house   Fig.2b: ravaged residential building

3. Many households have their household items were destroyed by fallen structures.
4. Some households also lost their Bananas cassava, and bean fields.
5. Mwamad Okello's zone had two goats that were killed by falling bricks. Over 120 chickens died and 5 households in Zone C on the landing site which lost up to 120 chickens per household.

Disaster management status after the effect

1. In the same landing site, some self-help projects like Prime Junior Primary School, a school with a total enrolment of 400 pupils as at the end of term I, 2015 was also severely devastated. Although at the time of field assessment, the management was trying to reinstate some of the fallen two building blocks (fig. 3); it was difficult to imagine the school would comfortably host all their second term pupil numbers. On this school, two buildings were deroofed and thrown down. This school which is private parent funded school which was seriously damaged but at the same time it is still at risk the way it was built (see the nature of its walls of some its building blocks in caption 1).

Caption 1: Nature of buildings that accommodate some of Uganda's children

From the field interviews among the affected households, most of the household heads had not attained education beyond Primary seven (table 2). This partly has contribution to the effects they are undergoing through. The survey did not compare unaffected households to these that fell culprit but my opinion is that more enlightened people have built permanent structures that survived this disaster.

Table 2: Highest education levels of HH heads of affected household

Highest education level	% age	Cumulative %age
Primary level	76	76
Secondary level	12	88
Tertiary level	0	88
Never studied	12	100

Fig. 3: Prime Junior School building block under reconstruction by 12/05/2015

Coping status	Percent	Cum. percent
Has started reconstruction	20	20
Has shifted to relative for accommodation	28	48
Has not yet reconstruction	16	64
Not affected	36	100

Recommendations

1. Government should respond to this community by putting mechanisms for reconstruction of this settlement;
    a. There are many vulnerable households (see list of affected <u>households</u>) whose livelihoods were completely devastated and these have requested for beddings and food stuffs in the short run.
    b. The health services department and other health service providers should respond to the affected community with mosquito nets, drug allocation to Bwondha Health Centre II to aid malaria and other disease control.

2. The National Forestry Authority and local government should offer the people along the shores tree seedlings and enforce the planting and management of the seedlings

along the shores to act as windbreakers for the fast moving wind that occasionally blows from the lake to the settlements.

3. The Forestry Landscape Restoration project under ministry of Water & Environment should take this opportunity to work with other agencies to restore green cover (tree stands) on hills namely Walujjo, Nairobi, Bwagu, Malindi and Lubango. The way nature reacts to its natural setting is very disastrous (it follows Newton's third law; which states that every action is met with an equal and opposite reaction). Therefore, the community development agents (CDOs, NGOs, Private Sector Organizations, etc.) should strategically package environment friendly information that will woo these communities to begin improving their own environment through replanting forest trees on these natural boundaries of the water body.

4. Bwondha landing site in itself is not well planned, there is no physical plan for this settlement yet it is rapidly growing into a very big town. Authorities should urgently act fast to develop a physically organized/planned settlement and use all available means including persuasive and authoritarian methods to reorganize the land site to international standards.

5. Building housing structures (whether for residence or commercial purposes) is a long term investment that requires time, money, scheduling, and quality management. When you look at all the structures that were demolished, it seems the affected households fell short of building requirements. Community development agents need to serious develop a comprehensive information package to sensitize communities to invest time and other resources including passing byelaws while building their housing structures. These will ultimate result in better housing structures that can resist forces of nature to a certain level possibly 90% acceptability. With this approach, it will in a long run reduce the cost incurred to the communities, government and other development partners.

6. There is need to strengthen the provision of education services (through schools, churches, mosques, community meetings, mass media, etc.) not only to Bwondha community but the entire district. This will go a long way to change the people's way of preparing for disasters relating to hailstorms, earthquakes, and the like by of

changing their perception on building housing structures as Matthew 7:24 – 27 advises, deforesting their own immediate environment. The educational institutions should emphasize practical exposures of pupils & students to the relationship between environment degradation and the negative effects on these very people's settlements. More generally they should redesign packages of information that bring together education, climate change management and improvement livelihoods. The existing schools that were affected (e.g. Prime Junior Primary school) or that are vulnerable should be prepared for managing such risks as well as the community in a long run.

References

UBOS, 2014. Uganda Census report 2014